Whispers of Clarity

By

Rosie Nora Miranda

Contents

Dedication

I share these experiences not only to honor my mentors but also to leave a record of the wisdom that has shaped my life—a personal legacy for my family and future generations.

It is my hope that, by sharing these influences, my family and future generations will continue to seek truth, growth, and a deeper understanding of who we truly are.

Acknowledgments

Since my early twenties, I have searched for mentors who truly understand how the mind works—those who can teach us how to shift our thinking to find greater clarity and improve every area of our lives. Over the years, I have been blessed with several remarkable teachers who have guided my journey.

I thank **José Silva** of Laredo, Texas—a self-taught electronics technician and the creator of The Silva Method. In the early 1970s, I attended a three-day seminar where we practiced techniques to expand the mind and awaken the sixth sense. His teachings, still recognized worldwide, continue to inspire those seeking to understand the true potential of the human mind. His work lives on through books, seminars, and resources available on YouTube.

I also thank **Dr. Joe Dispenza**, a neurologist and teacher whose work demonstrates the power of the mind and how we can consciously become the person we were meant to be. His lessons—shared through YouTube and other platforms—have profoundly enhanced my life. His decades of research and inspiration have reached millions around the world, and I am one of the many who have been transformed by his message.

Finally, I express my gratitude to the YouTube channel "**The Holy Spirit Within**." Although I do not know the creator's name, their teachings helped me awaken to the divine presence within—the God-given gift we all possess from birth. Through this awareness, I have learned to embrace and nurture the spiritual connection that guides and uplifts us all.

Chapter 1

Guided Meditations – A Journey Within

Meditation 1
Entering Stillness

Close your eyes. Feel the weight of the world loosen its grip. With every breath, you drift deeper into quiet waters, where thought no longer chases thought. Here, in the soft silence, you are safe. You are whole.

Meditation 2
Releasing the Noise

Imagine your body as a vessel. Every worry, every burden, like pebbles filling it. With each exhale, a pebble is lifted out, lightening your spirit. Soon, you are empty of heaviness, filled only with breath, with light, with peace.

Meditation 3
The River of Calm

There is a river inside you. Its waters flow slow, gentle, and eternal. When you rest your awareness here, the river cradles you. Every ripple whispers: you are enough, you are carried, you belong.

Meditation 4
Becoming the Sky

Look inward and see the sky. Vast, endless, unbroken. Your thoughts are clouds, drifting in and drifting out. You are not the clouds. You are the wide sky. Unmoved. Free. Eternal.

Meditation 5
A Sanctuary Within

Picture a door inside your chest. Behind it lies your sanctuary: a space built of warmth, of light, of quiet strength. Step inside. Feel the comfort of being home within yourself.

Meditation 6
The Breath of Renewal

Inhale deeply — a gift. Exhale fully — a release. With every cycle of breath, the old falls away, and the new blossoms. Your body knows. Your spirit knows. You are renewed

Meditation 7
Embracing Light

A golden light pours from above, spilling gently onto your crown, down your shoulders, into your heart. This light does not burn; it soothes. It fills the corners where fear has hidden. Now, only warmth remains.

Meditation 8
Resting in Presence

Here. Now. Not yesterday, not tomorrow. This breath. This stillness. Notice the miracle of simply being — your heart beating, your lungs breathing, life flowing. It is enough.

Meditation 9
The Garden of Peace

You walk through a garden, barefoot on soft grass. The air carries the fragrance of blossoms, sweet and pure. Every flower opens just for you, reminding you that beauty blooms when you are present to see it.

Meditation 10
Wholeness

Sit in your wholeness. Not a piece missing.
Not a flaw to be erased.
You, as you are, are complete. Feel the truth
of it echo in your body.
You are whole. You are home.

Chapter 2

Use God's Energy

Breathe.
Slow down.
And remember — the power that made the stars, the seas, the
mountains, is the same power that flows through you right now.

This is God's energy.

It is not distant. It is not far away.
It is here. It is alive. It is inside of you.

Every time you feel tired, uncertain, or overwhelmed… pause. Lift your heart upward, and open yourself to that eternal current. With every breath you take, His energy restores you.
With every step you make, His light strengthens you.

You don't walk in weakness. You walk in divine strength.

Let go of the doubts, let go of the fears, let go of the stories that tell you you're not enough.

God's energy does not run out. It is infinite. It is the fire in your soul, the peace in your heart, the light on your path.

Say this with me:
I am filled with God's energy.
I am guided by His light.
His strength flows through me now.

When challenges rise, you won't face them empty — you will face them charged with God's power.

When the world feels heavy, you won't carry it alone — His energy lifts you.

This is your reminder: Use God's energy.
Not later. Not someday.
Now.

Because it's already here.
It's already in you. It's already enough.

Chapter 3

The Whisper Within Recognizing the Voice of the Holy Spirit

There comes a moment in every believer's journey when silence becomes sacred. Not the silence of absence, but of presence — a holy stillness where words fall away and something deeper begins to speak. That something… is the Holy Spirit.

The Spirit does not always thunder or strike with lightning. Often, he whispers. He nudges. He breathes a gentle thought that carries the weight of eternity. Many mistakes whisper for their own conscience or imagination, but for those who have learned to listen, it is unmistakable, the soft yet steady voice of God dwelling within.

Jesus said in John 14:26: "But the Helper, the Holy Spirit, whom the Father will send in My name, He will teach you all things, and bring to your remembrance all that I said to you."

This is not merely comfort; it is clarity. The Spirit does not only console us in pain. He counsels us in confusion. He leads us through the chaos and into divine order.

Learning to Listen

The most difficult part about walking with the Spirit is not hearing Him — it's recognizing Him amidst the noise of the world. The world screams; the Spirit whispers. The enemy uses distraction, deception, and doubt to blur that voice, while the Spirit invites us into discernment, a place of inner stillness where we can sense God's direction even when the world is spinning.

When you begin to quiet your soul, when you begin to still the anxieties, the Spirit begins to move in ways that feel supernatural yet strangely familiar. He brings verses to mind you didn't know you remembered. He gives peace where logic demands panic. He leads you to people, to places, to purposes that align so perfectly it could only be divine orchestration.

You will know it's Him because His voice always aligns with His Word. The Spirit will never contradict Scripture. The Spirit's voice will never glorify fear, self, or sin. It always magnifies Jesus.

Walking in Step with the Spirit

The third chapter of your spiritual awakening begins when you stop asking, "Is this me, or is this God?" and start trusting that when your heart is surrendered, your instincts become sanctified.

Galatians 5:25 reminds us: *"Since we live by the Spirit, let us keep in step with the Spirit."*

That phrase — keep in step — is not passive. It means movement. Partnership. Harmony. To walk in step with the Spirit means to live in rhythm with Heaven, to allow your daily decisions to be directed by divine wisdom.

This is not always comfortable. The Spirit often leads us through the wilderness before He reveals the promise. He will take you places that stretch your faith, remove your comfort, and redefine your priorities. But every time you trust His leading, your spiritual hearing sharpens.

When the Spirit Speaks

The Holy Spirit speaks through:

- **Conviction,** not condemnation — He corrects without crushing.
- **Peace,** not pressure — when something is of God, there is rest even in uncertainty.
- **Confirmation,** not confusion — the Spirit will repeat a truth through different sources until your heart receives it.

When you learn to discern these patterns, your relationship with God shifts from a mere religion to a profound revelation. You begin to realize that prayer is not a monologue but a dialogue. You speak, He responds. You wait, He whispers.

The Power of Obedience

Hearing the Spirit is only half of the blessing. The true transformation happens when you obey. Every act of obedience deepens your sensitivity to His voice. Each time you respond to that gentle prompting to forgive, to step out, to speak up, or to remain silent, your spiritual muscles strengthen.

The Spirit does not force; He invites. But those who accept His invitations walk in power, peace, and divine purpose.

Romans 8:14 declares, *"For those who are led by the Spirit of God are the children of God."*

To be led means to follow. To follow means to trust. And to trust means to walk blindfolded with faith in the One who sees all.

Closing Reflection

The Holy Spirit is not a distant force or abstract theology. He is a living presence, the breath of

God residing in you. Every believer carries this divine spark, but few truly let it blaze.

When you surrender not only your will but also your ears, when you begin to listen with your heart, you realize the Spirit has been speaking all along.

He has been there in the quiet moments, in the restless nights, in the gentle reminders you thought were coincidences. He has been guiding, shaping, and whispering truth. The Holy Spirit is the voice within — the whisper of Heaven that leads you home.

Chapter 4

Think Clearly, Live Fully

Inspired by the full "Clear Thinking" series

Breathe. Slow down.
And remember—your mind is your compass.
When it's clear, you move with purpose. When it's clouded, you drift.

This is clear thinking.

It is not rigid. It is not cold.

It is alive. It is strong. It is yours.

Every time you feel overwhelmed, reactive, or lost in noise… pause.

Return to the center. Return to clarity.

With every breath, your thoughts settle. With every choice, your focus sharpens.

You don't think in chaos.
You think in calm.

Let go of the rush. Let go of the pressure. Let go of the belief that speed equals wisdom.

Clear thinking is not fast—it is focused.

It is the power to:

- Separate fact from feeling
- Question assumptions
- Recognize bias
- Listen with empathy
- Filter what matters
- Decide with courage

Say this with me:

I am a clear thinker. I am not ruled by emotion. I am guided by reason, grounded in truth, open to growth.

When confusion rises, you won't collapse—you will clarify. When opinions clash, you won't react—you will reflect.

This is your reminder: Use clear thinking. Not later. Not when it's convenient.

Now.

Because it's already within you, it's already strong. It's already enough.
And when you do—
You'll notice the shift.
Your words will carry weight. Your choices will align with your values.
Your relationships will deepen. Your path will become visible.

This is not just a skill. It is a way of living.

So when the moment comes—when the world feels loud, when your mind feels crowded, when your heart feels pulled—

Pause. Breathe.

And think clearly. Because clarity is not a luxury. It is your birthright.

It is the quiet strength behind every wise decision. It is the steady light in every storm. It is the calm voice that says, "Let's look again."

Clear thinking is not about being perfect. It's about being present. It's not about knowing everything. It's about knowing what matters. It's not about winning arguments. It's about understanding truth. So when you feel the fog creeping in—when emotions rise, when pressure builds, when the noise gets loud—

Don't fight it. Don't rush it. Just return. Return to the breath. Return to the facts. Return to the question: "What is really true here?"

Because in that moment, you reclaim your power.

You reclaim your clarity. You reclaim your mind.

And from that place, you move forward—not with fear, not with

confusion, but with quiet confidence.

This is your reminder: Think clearly.

Live fully.

And let your mind become the light that guides your life.

Chapter 5

The Whisper Within

I used to think the Holy Spirit was something distant, something reserved for saints, prophets, and people who seemed to have it all together. I thought it was a voice that came from the clouds, a wind that blew only in churches, a presence that hovered just out of reach.

But I was wrong.

The Holy Spirit is not distant. He is not a guest who visits only when we are perfect. He is not a reward for good behavior. He is the breath already in your lungs, the whisper already in your chest, the light already flickering in your soul.

He is within.

I didn't always know how to listen. I thought I had to strive, to perform, to earn the right to hear Him. I thought I had to be clean before I could be close. But the Spirit doesn't wait for your perfection. He meets you in your mess. He speaks in the silence between your thoughts, in the stillness beneath your striving.

He is not loud. He is not forceful. He is gentle.

He is the nudge that says, "Wait." The whisper that says, "Forgive." The peace that says, "You are not alone."

I remember the first time I truly heard Him, not with my ears, but with my spirit. I was sitting in the quiet, neither praying nor asking, just breathing. And in that stillness, I felt it: a warmth, a knowing, a presence that was not mine but somehow lived in me.

It didn't come with thunder. It came with tenderness.

And I wept not because I was sad, but because I realized I had never been alone. Not once. Not even in the darkest places.

The Holy Spirit had been there all along.

He had been there when I was lost. He had been there when I was angry. He had been there when I was numb. He had been there when I didn't believe.

He had been waiting not for me to be worthy, but for me to be willing.

Willing to listen. Willing to slow down. Willing to surrender.

And when I did, I discovered something I had never known: the Spirit is not just a guide, He is a friend. A comforter. A teacher. A well of wisdom that never runs dry.

He doesn't just point the way-He walks it with you.

He doesn't just convict you-He carries you. He doesn't just speak truth-He speaks love. And once you hear Him, you can't un-hear Him. Once you feel Him, you can't forget Him. Once you know Him, you can't unknow Him.

He becomes part of you.

So now, when the world gets loud, I go quiet. When the path gets dark, I look inward. When I don't know what to do, I listen.

Because the Holy Spirit is not out there somewhere.

He is here. He is within.

Finding the Holy Spirit within 2025

1. Find a quiet space, breathe deeply for 4 seconds, hold, and release. Repeat this exercise about 5 times, or until you feel completely relaxed.

2. Stay in the present, do NOT bring past experiences to the surface

3. Take one day at a time to begin with, do this breathing exercise often at the beginning until you are sure of this technique.

4. Once you are in a relaxed state of mind, begin to use your vision, especially as you are dozing off to sleep. Your unlimited subconscious mind is working on everything you projected for the next day.

5. This daily vision only takes 5 minutes in the morning and 5 minutes in the evening, especially as you are dozing off to sleep.

Doing this 10-minute exercise helps you get in touch with your Holy Spirit within, which is perfect and whole.

6. The next day, you wake up with a clear mind, no interruptions from the noises of the world, and you are connecting with your higher self.

7. As you repeat your daily breathing exercise, open your mind to bigger and better achievements. What do you want a year from now, ten years from now, and for the rest of your life, see it in your vision already complete, although you take it one day at a time.

www.ingramcontent.com/pod-product-compliance
Lightning Source LLC
LaVergne TN
LVHW070826220926
848486LV00027B/1
9781971610184